PID Programming Using RSLogix 500

By
Gary D. Anderson

Copyright 2015 by Gary D. Anderson
ISBN-13: 978-1523291588
ISBN-10: 1523291583

Table of Contents

Chapter 1: Introduction — Page: 1

Chapter 2: PID Control Loop — Page: 3

Chapter 3: PID Set-up & Programming — Page: 13

Chapter 4: Scaling for PID Instructions — Page: 23

Chapter 5: Basic PID Tuning — Page: 29

Chapter 6: Conclusion — Page: 39

Chapter 1: Introduction

In my previous books, "*Basic Concepts of PLC Programming*" and "*Advanced Programming Concepts*", I endeavored to show many of the foundational programming instructions and how they are commonly used in ladder logic programming. One topic covered was analog scaling and only a brief overview of the PID instruction. In this book I want to look into the PID instruction in greater detail and hopefully make its use in programming easier to understand and implement.

If you are unfamiliar with calculus concepts and terminology, PID programming and tuning can appear more challenging than many of the other tasks we normally accomplish with ladder programming. In reality, after we enter correct program variables as instruction parameters - the PLC processor does all the "heavy-lifting" as far as math calculations are concerned. However, the better we understand PID concepts and how terms come together to provide a control output, the easier it should be for us to tune and make changes when necessary. In fact, having a better grasp and understanding of PID control should make it easier to implement and program a control-loop regardless of what platform you may be using – PLC based or one of the many other programmable controllers currently used in industry. Many of these controllers have auto-tune features that work very well in many applications – still it is often necessary to fine tune some of the elements of the control algorithm.

In this book the main focus will be on the setup and use of the PID instruction that is part of the Rockwell Automation- RSLogix 500 instruction set. Before getting into that, we'll look into some of the more general terms and math concepts pertaining to PID control loops.

Some of the difficulty in understanding PID control is dealing with the whole "time" element involved with calculating integral and derivative values. When programming PID instructions into ladder programming we're not only dealing with the scan time of the program, but also several variables that we must set, such as the loop update time parameter, and the setup of the timing mode - STI or Timed. Also the dynamics of the process itself introduce factors that determine how well it can be controlled. Process dynamics are the physical forces that affect a process – whether it's the axis movement of a robotic arm, maintaining a specific liquid level in a tank, or raising a process to a desired temperature. These physical forces, when talking in PID language are usually referred to as "disturbances", and are the reason that - while two processes may be similar to one another – they are never entirely identical and each must be individually tuned to produce the best results.

I've heard the saying over the years, that tuning a PID control loop is more "art than science". While I'm sure this is somewhat true – a better understanding of the science involved will at least allow us the opportunity to begin working on our art project. I have devoted a chapter to tuning considerations – again focusing on usage within the RSLogix programming instruction set.

As technicians, I believe the study of control theory is a valuable and worthwhile endeavor – certainly there are college level courses devoted to all aspects of control programming. Control loops can be programmed using C language or even actual mathematical coding within the ladder logic; however, Allen Bradley has provided us with what is basically a small macro program – the PID instruction. In the majority of cases it is more than sufficient provided we take the time to get the scaling for input and output variables correct and the PID properly tuned for the process application.

Chapter 2: PID Control Loop

It's impossible to understand a PID controller without some basic understanding of the math it uses to produce an output signal. It is after all, an algorithm that takes a measured value, does comparison with a desired set point value - and then uses math calculations to produce an output signal. All this is done in a decidedly "non-linear" fashion with changes in our control variable constantly being altered with every program scan of the PID program or instruction. Essentially PID is an acronym for three mathematical terms: *proportional*, *integral*, and *derivative*, that are calculated and summed together to form an output signal.

Here is the basic schematic depicting a typical PID control loop. The object of the PID controller is to produce a control variable that will be used to provide an analog output voltage or current signal - which can then be used to <u>reduce error</u> between the set point and process variable.

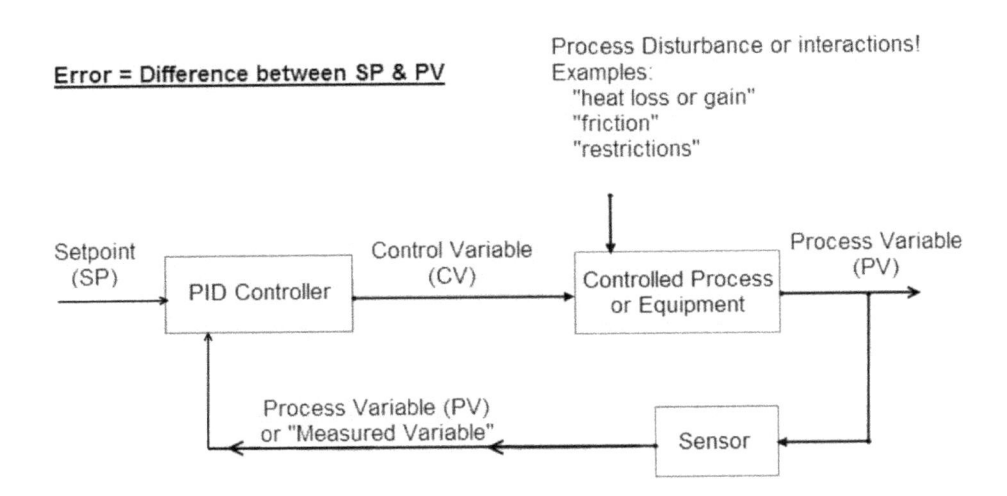

Error = Difference between SP & PV

Process Disturbance or interactions!
Examples:
"heat loss or gain"
"friction"
"restrictions"

Setpoint (SP) → PID Controller → Control Variable (CV) → Controlled Process or Equipment → Process Variable (PV)

Process Variable (PV) or "Measured Variable" ← Sensor

Another way of looking at a control loop is in the following format. This may be more descriptive in showing the relationship between calculated values and their impact on the control variable (CV). Note that I've shown error (E) as equal to SP minus PV, but in any controller I've used, this can also be set for PV minus SP in the setup menu - depending on the control action you desire.

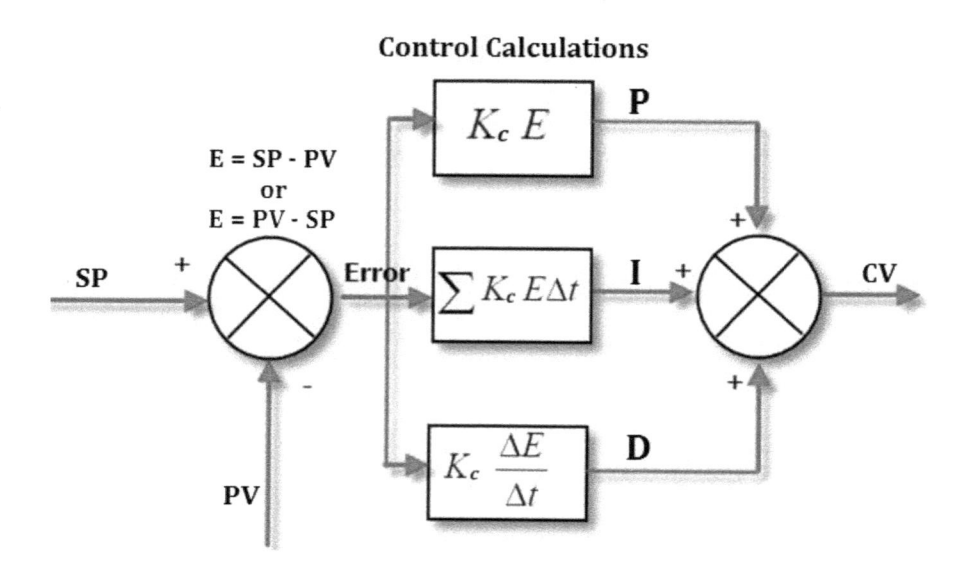

Control Calculations

$$E = SP - PV \text{ or } E = PV - SP$$

Blocks: $K_c E$ (P), $\sum K_c E \Delta t$ (I), $K_c \dfrac{\Delta E}{\Delta t}$ (D)

Inputs: SP, PV → Error → P, I, D → CV

The component algorithms can be used together or separately. I've worked with many applications where only the proportional element was used. P & I controllers are commonly used for various process controls, and P & D control for servomotor and positioning applications such as CNC machines or robotics. The important point to remember is that each of the three terms of the algorithm, make their own unique contribution to the controller's output. *Proportional* control is used to rapidly reach for a setpoint value; *integral* control to correct a smaller error value that proportional alone can't correct without oscillation or overshoot, adding the *derivative* element can further stabilize and prevent overshoot in the process variable.

Here is the general *dependent-gains* equation used in the SLC 500 instruction set for the PID calculation. Note that the proportional gain (Kc) is a factor for all three terms – therefore the name, since the output from any of the three terms is dependent on K_C.

$$Output = K_C\left[(E) + \frac{1}{T_I}\int (E)dt + T_D \cdot \frac{\Delta E}{\Delta t}\right] + \text{Feed Forward/Bias}$$

Proportional (P) control adjusts the output signal based on the magnitude of the error at the current time. As you can see from the following equation, the calculated output value is <u>directly proportional</u> to both the size of the "error" and also the value you have entered as gain (Kc). The larger the error at any instant, the greater the output signal that is produced as the control variable.

$$Output = K_C(E)$$

Almost any control loop system will rely on proportional control for the main part of its corrective action. I think of it as the baseball player that always swings for the home-run. When used alone it will generally cause a process variable to settle into a condition where a steady state error (difference between SP and PV) is maintained. If gain is set too high the output and PV will oscillate and become unstable.

Here is a typical example of a control loop using only the proportional algorithm to reach and maintain a temperature set point. Oscillation can be a great deal more dramatic than what I've shown in these diagrams and can become completely unstable as the proportional control attempts to provide control output that will settle the constantly occurring error.

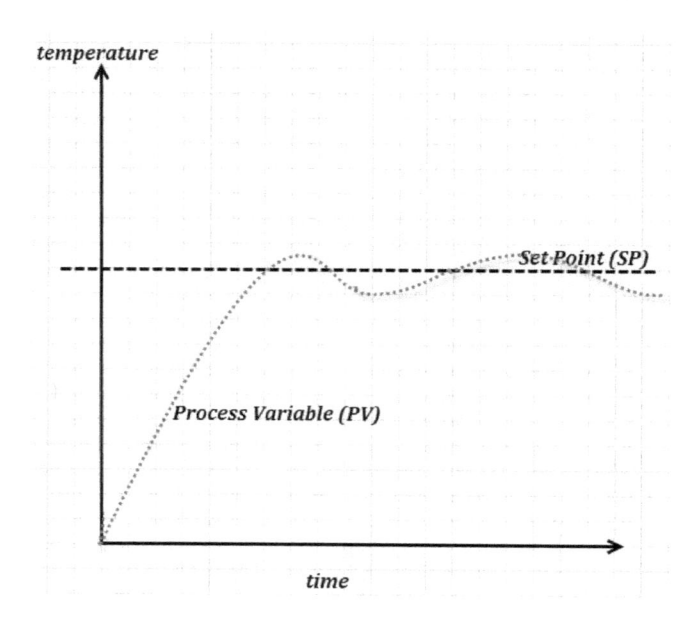

[Graph: temperature vs time, showing Process Variable (PV) rising to oscillate around Set Point (SP)]

Another possibility might be a graph that looks similar to the one shown in the following example. Disturbances to the PV, such as the addition of raw material into a process or some other type of heat loss, contribute to a "steady-state" error condition that the proportional control alone cannot entirely correct. Other factors that contribute to steady-state error can be friction, restriction to flow, or even the weight of a machine component.

Some degree of error continues to exist using proportional control because at some point, <u>where calculated error become very small</u> – the output voltage from the controller also becomes extremely small. With the error almost entirely corrected, the output voltage to control devices may be not enough to accomplish any further corrective action in terms of valve or motor movement. Because of the physical dynamics we've mentioned, this steady-state error, sometimes referred to as offset, remains. With some processes this might be acceptable - while in others, such as servo positioning control, it would not.

temperature

.... set point (SP) value

Process Variable (PV)

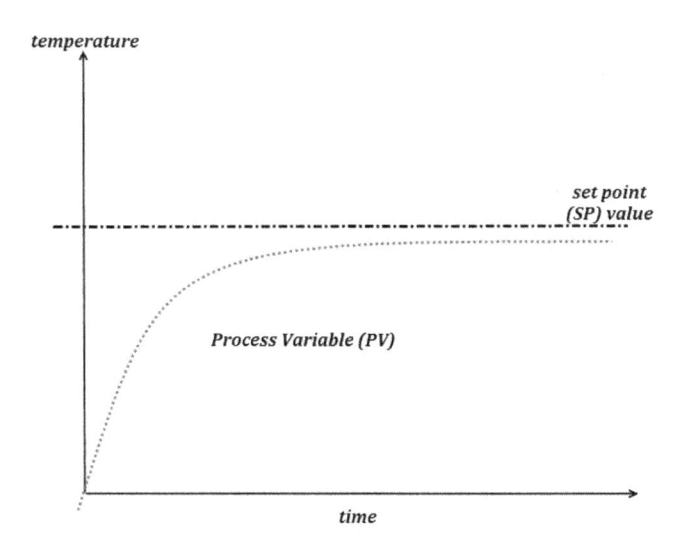

time

Proportional control makes a large and dramatic contribution in bringing a measured process variable closer to a desired set point value. Once reaching the point where a steady-state error exists, something more is needed to overcome the remaining small margin of error. The answer to this problem is the use of "integral" and possibly the "derivative" components of the PID calculation.

Integral (I) control provides an output proportional to the <u>accumulated error</u> over past time segments. More accumulated error over the selected time period produces a larger control signal that is summed into the CV. This in turn produces a larger output to devices that directly control the process. To say it another way - if an error or "offset" still remains after proportional control has done its job, then the accumulated error signal of the integral becomes large enough to further move the process and eliminate remaining error. This is the equation for the integral term as used by Allen Bradley with the RSLogix 500 application.

$$Output = K_C \cdot \frac{1}{T_I} \int (E) dt$$

The T_I term essentially establishes the length of the time segments, I think of it as "resolution" within the overall timeframe set by the loop-update time setting. An error value, however small, is calculated for each time-segment and accumulates into a summed value. This value is applied to the control variable output as the PID is executed during the loop-update timed cycle.

In many applications PI is used to reach and maintain a specific set point. When properly tuned, adding the integral component to the control variable can bring a process variable through the steady state error range without undue overshoot or oscillation, as in the example graph that follows.

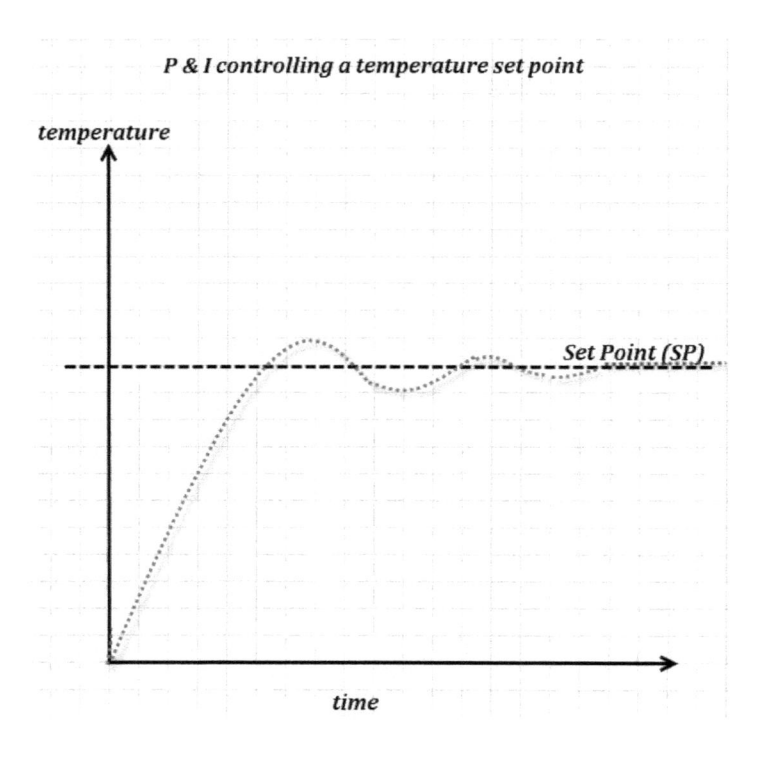

P & I controlling a temperature set point

One of the most important concepts to understand for setting integral action is the time variable. This setting is often referred to as "reset", and as the term implies, is derived from the way the integral periodically adds to the controller output by repeating the proportional action - in this case the T_I term determines the "minutes-per-repeat" that the integral takes to repeat the proportional action. Integral control deals with the reset variable in one of two methods, either using "reset rate" or "reset time". While these are presented differently in equation form, they ultimately accomplish the same function establishing the length of time in which the calculated error is accumulated. Reset Rate can be viewed as the number of "repeats per minute", while Reset Time denotes the "minutes per repeat"- so they are the reciprocal of one another.

The focus in this book series is the RSLogix 500 application software and the SLC 500 and MicroLogix family of processors. These utilize the "minutes per repeat" algorithm for the integral calculation. Later model Allen Bradley processors, such as the ControlLogix/CompactLogix processors; also make available the "repeats per minute" type equation to derive output values. This is an important difference to be aware of when preforming initial setup and tuning. When using the *minutes-per-repeat* model - the larger we set the reset value term (T_I) – the smaller the integral action becomes. This is because the integral action is spread over a longer time period.

Conversely, a smaller setting for the (T_I) value will cause a larger integral action. The smaller reset setting is calling for the control to repeat the proportional action – but to do so in a shorter timeframe. This concept might be more clearly understood when you look at the actual equation used by the SLC 500 processor – simply notice that the integral term (T_I) is used as the denominator of the fraction. Therefore the larger the value for Ti – the smaller the factor value that multiples the accumulated error.

Derivative (D) control adjusts the control signal in proportion to the rate-of-change of error with respect to time. In contrasting this with the other two control elements consider the following. The "proportional" element acts upon the "instantaneous" or current error. The "integral" value acts upon accumulated or "past" error. The "derivative" term produces an output value based on anticipated or future error - given its current rate of change. As such, the derivative is used to "dampen" and stabilize the overshoot and oscillations that may be presented when using P or PI, and to more quickly bring a process to a desired set point.

Derivative is sometimes simply referred to as "rate" and shown in our equation as:

$$Output = K_c \left[T_D \cdot \frac{\Delta E}{\Delta t} \right]$$

Note that the PID instruction setup menu gives two options for <u>derivative action</u> – the one shown in the above equation, or the rate-of-change in the PV over time as shown below. This option is selected by the (DA) control bit in the setup menu – we will cover most of these control block settings in the following chapter.

$$Output = K_c \left[T_D \frac{\Delta PV}{\Delta t} \right]$$

This of course is the same basic equation used to calculate "slope" or the "m" value of a linear equation of the form y = mx + b and graphed using Cartesian XY coordinates - the x-axis representing "time" and the y-axis representing the "error".

The rate-of-change in error with respect to time is then multiplied by the derivative time factor T_D, and by whatever proportional gain we have set for our control system. Since the RSLogix 500 instruction uses the dependent-gains model, remember that all three equations are factored by the same proportional gain value (K_C).

In the following chapter, I'll recap some of the terms we've covered so far and introduce a few new ones that are used when actually programming PID control with RSLogix 500. I'll also be showing setup screens, integer data file examples and ladder programming associated with PID ladder logic – so we'll revisit these terms and concepts again - in greater detail.

As a side note, RSLogix 5000 and the ControlLogix platform go beyond the standard PID instruction as used with SLC 500 processors. The RSLogix 5000 instruction set includes the PIDE (enhanced PID) instruction, which uses a velocity algorithm based on the change-in-error rather than actual measured error in its proportional term calculation. It can use the dependent-gains form of the equation and also the independent-gains model, which makes it possible to set different gain values for each of the three terms comprising a PID control loop. Some of the other advantages of the PIDE include the ability for adjusting these gain values in a running loop – still in automatic mode, whereas the positional algorithm used with the RSLogix 500 software requires switching from auto to manual and creates a "bump" in the output variable. While I'm sure there are many advantages in using newer software and controllers - I still find RSLogix 500 and SLC 500 controllers widely used – and I believe will be for a long time to come.

Once again, if you understand the foundational math concepts, terms, and variables used with the RSLogix 500 and the SLC 500 platform – then programming and tuning should be easier with any other controller you may use. I've devoted a later chapter to the procedure of tuning using the Zeigler-Nichols method. While this will be shown using the PID instruction parameters that are part of the RSLogix 500 instruction set, the same method applies to any number of other controllers such as the stand-alone models shown below.

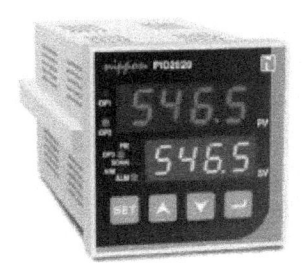

Chapter 3: PID Set-up & Programming

Here is a typical PID instruction used in a ladder logic program. All-in-all it looks pretty innocuous and is relatively easy to use once you understand some of the basic elements of PID control. Going at it in a systematic way is always helpful – at least for me, and it's often a good practice to simply make some notes and sketches that outline what you want to accomplish by using a PID instruction. This can be particularly true of any scaling required for the input (PV), setpoint (SP), and the controller output (CV).

In terms of memory allocation and set-up, the PID uses twenty-three words of memory within an integer data file. Rather than using the default N7 integer file for this control block, it is often less confusing to set up a separate integer file, such as N10 or N11 for PID usage. This is the case shown in the following example.

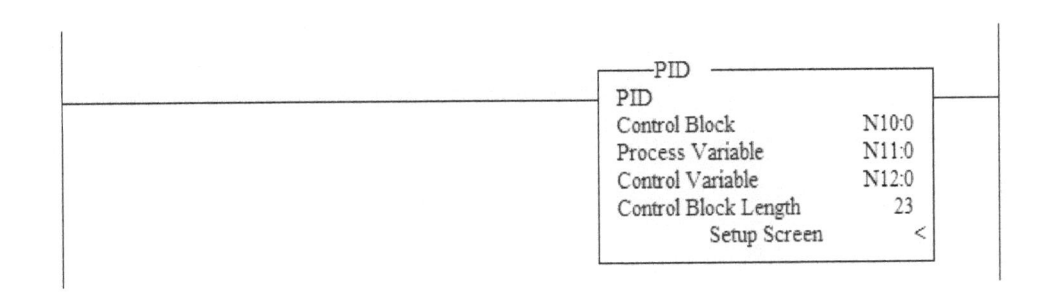

```
                          ──── PID ────
                         PID
                         Control Block          N10:0
                         Process Variable       N11:0
                         Control Variable       N12:0
                         Control Block Length      23
                                  Setup Screen      <
```

As indicated, the PID instruction uses 23 words of dedicated memory for its different control elements. I will be covering these in more detail, but if you were to simply double-click on the N10 data-file in the ladder application project tree, the information contained in the control block would look something like this.

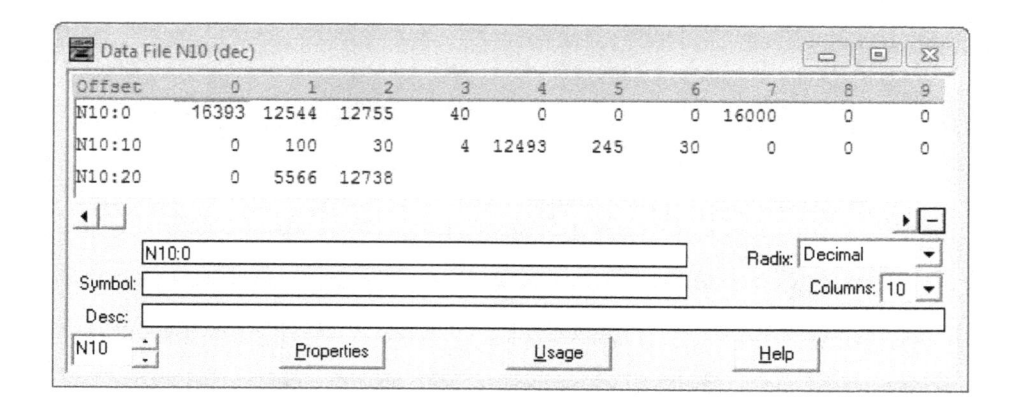

```
Data File N10 (dec)
Offset        0      1      2     3     4     5     6      7     8    9
N10:0      16393  12544  12755   40     0     0     0   16000    0    0
N10:10         0    100     30    4  12493   245   30       0    0    0
N10:20         0   5566  12738
```

As you can see, N10/word 0 or N10.0 begins the control block which ends with word N10.22. Remember that each word element contains 16 bits so we're looking at a combined total of 368 bits of allocated memory per PID instruction.

Also be aware of the fact that while the PID *instruction* contains addressing for the process and control variables, these addresses are "user-defined" and separate from the 23 word control block. These are simply telling the PID where to look for its input (PV), and also where to send the resulting output value (CV). These input and output components of the PID usually require scaling. The PID instruction will accept and produce integer values that range from 0 to 16383 for both input and output. Setpoint input can also be scaled in the range of 0 to 16383, or to the values you enter for *Smax* and *Smin* in the setup menu screen, these are words 7 and 8 in the control block. The *Smax* values can range from (-16382 to 16383) for 5/02 processors, and (-32767 to 32767) for 5/03's and later. The *Smin* ranges are (-16383 to 16382) and (-32768 to 32766) respectively.

Clicking on the "Setup Screen" at the bottom of the instruction will bring up the following menu which allows you to program the many elements and to tune the control loop. Since we're not delving into tuning just yet, the following example is using "proportional" control only – notice that the T_I and T_D terms are both set to zero. This basically takes them out of the control loop where the only part of the equation contributing to the CV is the proportional element: ($K_C * Error$).

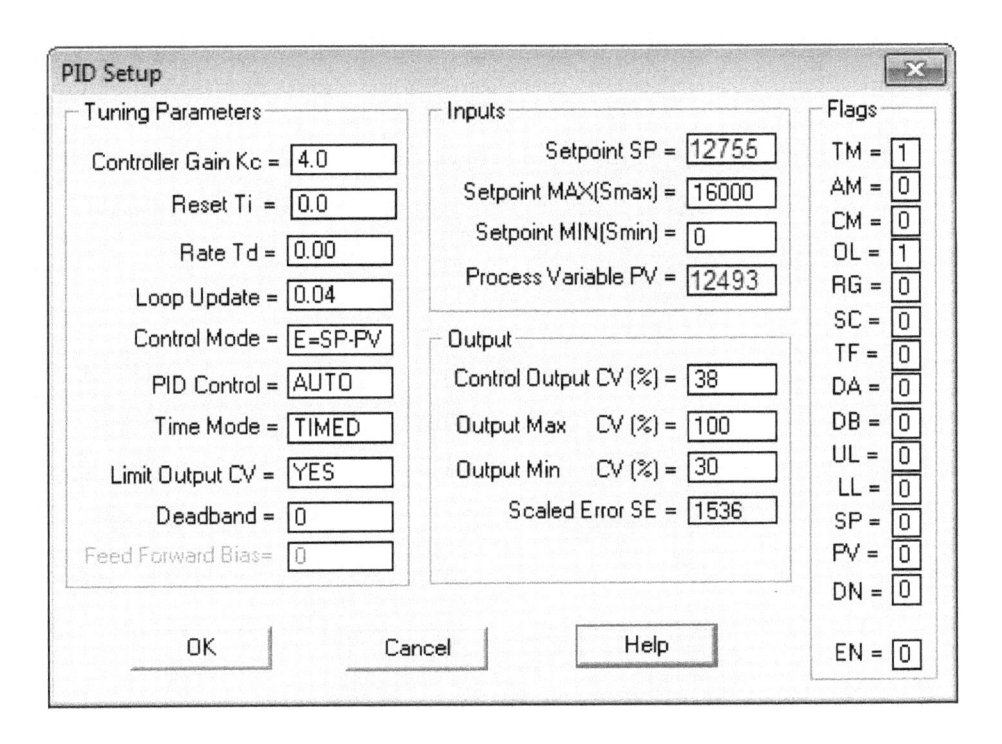

```
PID Setup
┌─ Tuning Parameters ──────────────┐  ┌─ Inputs ─────────────────────────┐  ┌─ Flags ──┐
│  Controller Gain Kc = [4.0]      │  │        Setpoint SP = [12755]     │  │ TM = [1] │
│        Reset Ti = [0.0]          │  │  Setpoint MAX(Smax) = [16000]    │  │ AM = [0] │
│         Rate Td = [0.00]         │  │  Setpoint MIN(Smin) = [0]        │  │ CM = [0] │
│     Loop Update = [0.04]         │  │  Process Variable PV = [12493]   │  │ OL = [1] │
│    Control Mode = [E=SP-PV]      │  ├─ Output ─────────────────────────┤  │ RG = [0] │
│      PID Control = [AUTO]        │  │  Control Output CV (%) = [38]    │  │ SC = [0] │
│       Time Mode = [TIMED]        │  │  Output Max   CV (%) = [100]     │  │ TF = [0] │
│   Limit Output CV = [YES]        │  │  Output Min   CV (%) = [30]      │  │ DA = [0] │
│        Deadband = [0]            │  │    Scaled Error SE = [1536]      │  │ DB = [0] │
│   Feed Forward Bias= [0]         │  └──────────────────────────────────┘  │ UL = [0] │
│                                  │                                        │ LL = [0] │
│                                  │                                        │ SP = [0] │
│                                  │                                        │ PV = [0] │
│                                  │                                        │ DN = [0] │
│       OK          Cancel              Help                                 │ EN = [0] │
└──────────────────────────────────────────────────────────────────────────────────────┘
```

Here you can see just about everything needed to setup a PID instruction in a ladder program and tune it for the process it controls. In this example, our proportional gain (K_C) is set at 4.0 and other important details such as Loop Update and Control Mode can be viewed and changed. On the right side of the setup menu – under the "Flags" heading, are most of the discrete bits that comprise word 0 of the control block. Some of these are indeed only "flags" which I think of as "read-only" while others can be written to or toggled. For instance if you change bit 0 (TM) from "1" to "0", you will see the Time Mode option change to "STI" from its current selection of "Timed".

On the following pages we'll look into the settings and parameters that are specific to the Allen Bradley SLC-500 family of controllers, and many of the general terms that pertain to PID control and tuning – regardless of the controller used.

Terms & Settings

Kc: Proportional controller gain. Applied to all three algorithms with the dependent gains model, and is usually set to approximately ½ the value of the setting that will cause oscillation in the output when Ti and Td are set to zero. The Kc proportional gain value resides in word 3 of the control block and ranges from 0 to 3276.7 (when RG= 0), or 0 to 327.67 (when RG = 1). RG is a discrete bit setting and part of word 0 in the control block – more on this later in the chapter.

Feed Forward: FF is basically an open-loop element, such as a predetermined offset value set by the operator, which combines with the PID (closed-loop) output. The PID compensates for whatever error difference remains between the SP and the PV <u>after</u> applying the Feed Forward value. Although not shown on the setup screen it is addressed to word 6 of the control block and can be used to add a desired value that may offset disturbances within the process.

E: The "error" equal to the difference between the SP and PV. This can be selected as either (E=SP-PV) or as (E=PV-SP) depending on the type of operation under control such as a "heating" application versus a "cooling" application.

Process Variable (PV): This is sometimes referred to as the *measured* variable. It is the analog input which provides a snapshot-in-time of the process you want to control. A few examples could be the temperature of an oven, maintaining a specific liquid level in a tank, a flow rate from a pump, or the axis position of machine. The PV analog input scaled range is 0 to 16383, and is stored as an integer value.

Setpoint (SP): The setpoint is the value established and entered by the operator or process program. This is the desired control value you want the process variable to reach and be maintained at - until changed by the operator or program.

Control Variable (CV): This is the output value produced by the PID control algorithm which will translate – by an output module D/A conversion - into some form of output signal that moves the process toward the desired SP. The CV output value produced by the PID instruction will range from 0 to 16383 and can be further scaled for your desired I/O and application.

T_I or Reset: Also called integral gain and is generally set equal to the natural period as measured using the proportional gain alone. This is the "reset time" for the integral equation and is one of the parameters that must be set when programming the PID instruction. If set to zero, the integral doesn't add anything to the CV output. The T_I value resides in word 4 of the PID control block.

T_D or Rate: Also called derivative gain and generally set to a value of approximately one-eighth of the Reset Ti integral time value. This provides an additional factor that, together with K_C multiplies the calculated rate-of-change in either the (E) or (PV), depending on which derivative action is selected. Of course if T_D is set to zero, just like the T_I reset, the overall product would then be zero and so would not contribute any useable output value.

Mode (TM): This bit specifies when the PID is in timed mode (bit set to 1) or STI mode (bit set to 0). This bit can be set or cleared by instructions in your ladder program. When set for timed mode, the PID executes and updates the CV at the rate specified in the loop update parameter (word 13). If TM is set for STI mode, which is an acronym for "sequence timed interrupt", the PID executes and updates the CV every time the PID instruction is scanned. If the STI option is selected then the PID instruction should be programmed within an STI interrupt subroutine. This subroutine file number is designated in word S:31 of the status file. If this word is set to zero then the STI is disabled. The STI routine, when used, should have a time interval equal to the setting of the PID "loop update" parameter. This STI time interval is also set in the status file, word S:30. If S:30 is set to zero, this too disables the STI.

Timed mode should be used if the scan time of the processor is at least 10 times faster than the loop update time. Loop update time, in the preceding example, is set at 0.04 seconds on the setup menu. While not shown, the average scan time for this processor is at a 3 milliseconds. So 10 times our scan time would be equal to 0.030 seconds – well within the selected 0.040 loop update time. For this reason I'm using the timed mode in this example application.

Another consideration when specifying mode of operation is the relationship between program scan time and the natural period of oscillation of the loop. General consensus is that if the "natural period" of the loop - the time between peaks when you have a PID loop with only the "proportional" algorithm in use, is greater (slower) than 10 times the scan time as shown in the status file, then used timed mode. If it's less than 10 times the scan time - then consider using STI mode.

Once again, a general rule for PID setup:

Use **Timed Mode** if Natural Period $_{(Output)}$ is (10 x) greater than scan time.

Use **STI Mode** if Natural Period $_{(Output)}$ is less than (10 x) scan time.

Loop Update: The loop update (word 13) is the time interval between PID calculations. Remember from the previous description that, when in timed mode, the PID updates output per the loop-update time setting. The entry for this setting is in 0.01 second intervals. It is usual practice to set loop update time five to ten times faster than the *natural period* of the load - as determined by setting the reset and rate parameters to zero and increasing the proportional gain until the output begins to oscillate. The natural period is the time it takes the output to make one cycle – such as a peak-to-peak time measurement. When in STI mode, the loop update value must equal the STI time interval value entered in status word S:30. The valid range is 1 to 1024 intervals.

Deadband: The deadband extends above and below the set-point by the value entered. The deadband is entered at the zero crossing of the process variable and the set-point. This means that the deadband is in effect only after the process variable enters the deadband *and* passes through the set point. The dead band represents the range over which the primary controller will allow the process variable (PV) to deviate without exerting any correction. As you can see, the proper setting of a deadband range will prevent short-cycling that could present a problem in some types of process applications. The valid range is 0 to the scaled maximum or (0 to 16,383). The flag bit 8 in word 0 is set whenever the PV is within the zero-crossing deadband range.

Scaled error: The difference between the process variable and the set-point scaled values.

Auto/Manual (AM): The auto/manual bit can be set or cleared by instructions in your ladder program. When off 0, it specifies automatic operation. When on 1, it specifies manual operation. In automatic operation, the instruction controls the control variable (CV). In manual operation, the user manually controls the CV.

Control Mode (CM): Selects either forward or reverse action, and so toggles between two methods of determining error: (E=PV-SP) for forward or direct acting, and (E=SP-PV) for reverse acting. When set to 1, *forward acting* (E=PV-SP) causes the control variable to increase when the process variable is greater than the setpoint. When set to 0, *reverse acting* (E=SP-PV) causes the control variable to increase when the process variable is less than the setpoint.

Derivative Action (DA): When this bit is set to 1, the derivative (rate) action is evaluated on the error (E) instead of the process variable (PV). When cleared (set to 0), this bit allows the derivative to be evaluated based on rate-of-change of the PV.

As you can see, some of these terms are set directly from the PID setup menu. Other parameters, such as the process and control variables, are simply entered by address to the main PID instruction from within the ladder logic. These addresses hold integer values that also require scaling for the analog input (PV) and the analog output (CV) because the PID instruction deals only with numeric values in the range of 0 to 16383. For this reason other instructions such as the SCL are used and also often some "qualifier" conditions which can prevent a processor fault if an overflow occurs during execution of the PID instruction.

Before getting into those issues – here are two charts which give a breakdown of the 23 words allocated to a PID control block. Note that the bits in word 0 are discrete bits, while the remaining 22 words of the control block hold analog data. As you can tell from the following example, many of the bits and word values used by the PID instruction are *read-only* or listed for *internal use* while others are selected during setup and configuration of a PID instruction.

Here are charts describing the discrete bits of word 0 and simple descriptions for the remaining 22 words of the control block with setup options.

\multicolumn{4}{c}{CONTROL BLOCK FOR PID / WORD 0 Discrete Bits}
Bit	Symbol	Description	Setup Choices
0	TM	Time Mode	Timed or STI
1	AM	Auto / Manual	Auto or Manual
2	CM	Control Mode	E=SP-PV or E=PV-SP
3	OL	Output Limiting	NO / YES
4	RG	Range Enhance	0, 1
5	SC	SP Scaling	(Read only)
6	TF	Update to Fast	(Read only)
7	DA	Rate Action	0, 1
8	DB	PV is in Deadband	(Read only)
9	UL	CV Upper Alarm	(Read only)
10	LL	CV Lower Alarm	(Read only)
11	SP	SP Out-of-Range	(Read only)
12	PV	PV Out-of-Range	(Read only)
13	DN	PID Done	(Read only)
14	RA	Rational Approx	
15	EN	PID Enabled	(Read only)

Word	Symbol	Description
0	Discrete Bits	Setup and Read-only bits
1		Sub Error Code
2	SP	Setpoint
3	Kc	Proportional Gain
4	Ti	Reset (Integral)
5	Td	Rate (Derivative)
6		Feed Forward Bias
7	Smax	Setpoint Max
8	Smin	Setpoint Min
9		Deadband
10	internal use	Do not change
11		CV Output Max%
12		CV Output Min %
13		Loop Update Time
14	PV (scaled)	SPV (Scaled Process Variable)
15	SE	Scaled Error
16	CV%	Output CV percent
17	internal use	Do not change
18	internal use	Do not change
19	internal use	Do not change
20	internal use	Do not change
21	internal use	Do not change
22	internal use	Do not change

PID Control Block - 23 Word

As you can see, these terms are set from within the setup menu, however many can also be used within the ladder program by other instructions. Examples of this might be the usage of the PID "done" bit or the upper and lower alarm bits for the control variable. Thus the different bits of word 0 establish the characteristics of how you want the PID to operate, but also can serve other functions within the program logic.. Looking once again at the N10 datafile and changing the view to binary allows us to see exactly which flag bits are set in this particular PID application.

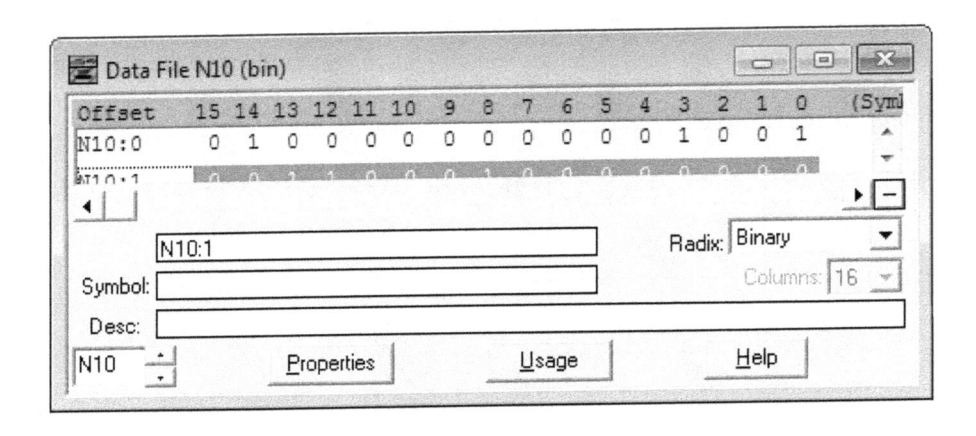

In this illustration, you can see that bits 1, 3, and 14 of word 0 are all set to 1. These "flags" are indicating that *timed mode* and *output limiting* have been selected. These parameters tell the processor how often to calculate and update PID output and also to limit CV output to the limits set by words 11 & 12 of the control block. These words; *Output Max%* and *Output Min%* on the setup menu, establish the allowable range for the control variable, in my example given as percentages.

Another bit that is set on this processor is bit 14, the RA or *Rational Approximation* bit. This option is available with certain firmware revisions and is available with this application. The instruction set manual simply states that having this bit set produces a calculated output that has greater accuracy. It is a bit that doesn't appear on the setup menu but can be set in the datafile (N:10.0/14) if desired.

It's important to give careful consideration in the use of these bits and word values, especially if writing and implementing a new program. As you can see, some of these are dynamic, and so are set to 1 at various times during PID execution while at other times reset or cleared. Others bits and values you set when programming the PID are static and remain unchanged during the program scan. Complete descriptions can be found in the RSLogix 500 Instruction Set manual for each of the control bits and also for the different word elements used within the PID control block.

Chapter 4: Scaling for PID Instructions

For the sample program used to illustrate PID programming in this book, the following rung of ladder logic is shown along with its setup menu.

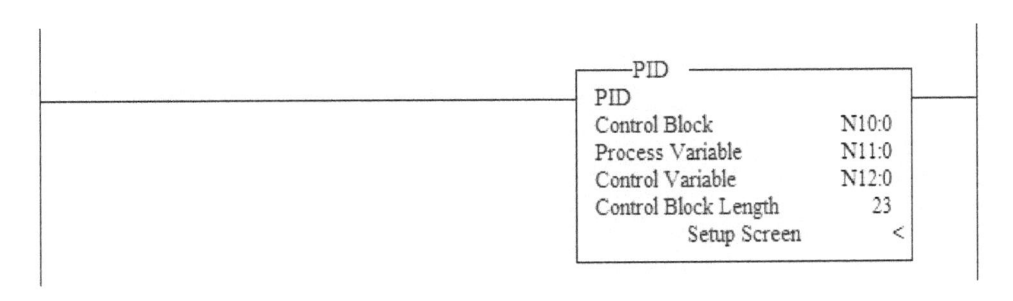

```
                              ┌─PID──────────────────────────┐
                              │ PID                          │
                              │ Control Block        N10:0   │
                              │ Process Variable     N11:0   │
                              │ Control Variable     N12:0   │
                              │ Control Block Length   23    │
                              │           Setup Screen    <  │
                              └──────────────────────────────┘
```

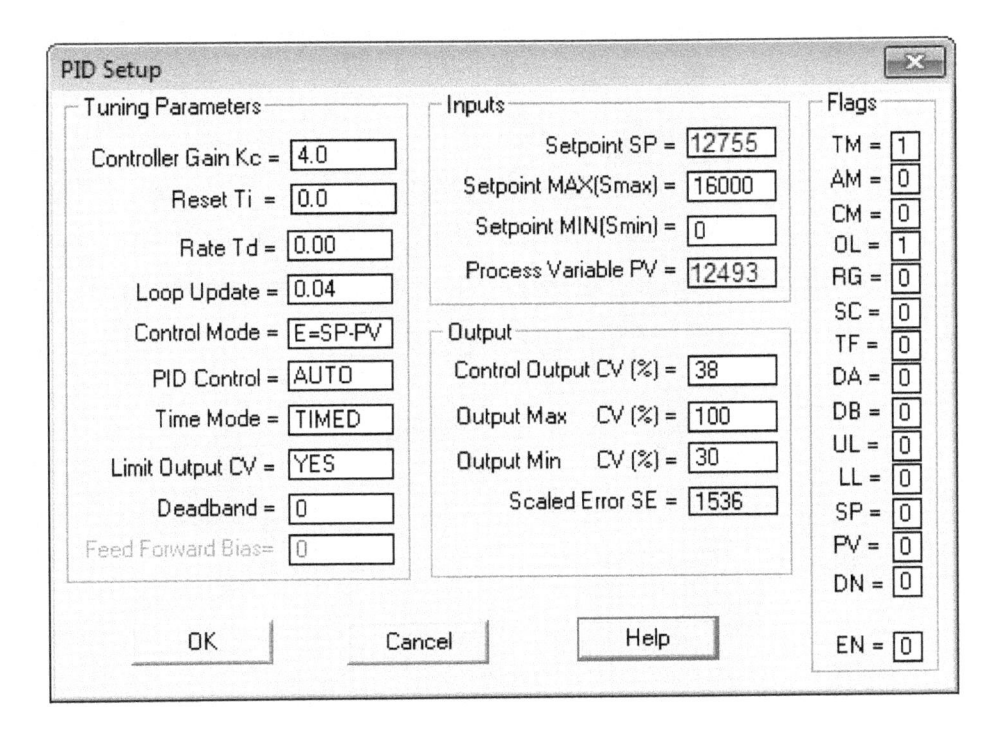

PID Setup

Tuning Parameters
- Controller Gain Kc = 4.0
- Reset Ti = 0.0
- Rate Td = 0.00
- Loop Update = 0.04
- Control Mode = E=SP-PV
- PID Control = AUTO
- Time Mode = TIMED
- Limit Output CV = YES
- Deadband = 0
- Feed Forward Bias = 0

Inputs
- Setpoint SP = 12755
- Setpoint MAX (Smax) = 16000
- Setpoint MIN (Smin) = 0
- Process Variable PV = 12493

Output
- Control Output CV (%) = 38
- Output Max CV (%) = 100
- Output Min CV (%) = 30
- Scaled Error SE = 1536

Flags
- TM = 1
- AM = 0
- CM = 0
- OL = 1
- RG = 0
- SC = 0
- TF = 0
- DA = 0
- DB = 0
- UL = 0
- LL = 0
- SP = 0
- PV = 0
- DN = 0
- EN = 0

OK   Cancel   Help

As you can see, the control block is addressed to integer file N10:0, and as we've discussed, uses words 0 through 22 for all the different control and calculation elements that it will perform. Our PID instruction also specifies the address of our process and control variables, PV and CV, which are not directly a part of the control block. An element that does reside in a control block address is the setpoint. As you can see from the control block structure chart, the SP resides in word 2. Whatever means are used to input the setpoint, in this case an HMI, the instruction will look in word N10:2 for this value.

The following examples show how these three components, the SP, PV, and the CV are scaled to the range required by the PID instruction and needed by the application – in this case a process using only proportional control of a hydraulic ram.

Setpoint Scaling:

Here is the program logic for the setpoint (SP) along with an illustration detailing the scaling process. In all of these examples I've used the SCL instruction, but of course straight-up math instructions can be used or the SCP instruction if supported by your processor. Using the SCL, once again, gives the opportunity to go through the steps that best explain the whole concept of analog scaling.

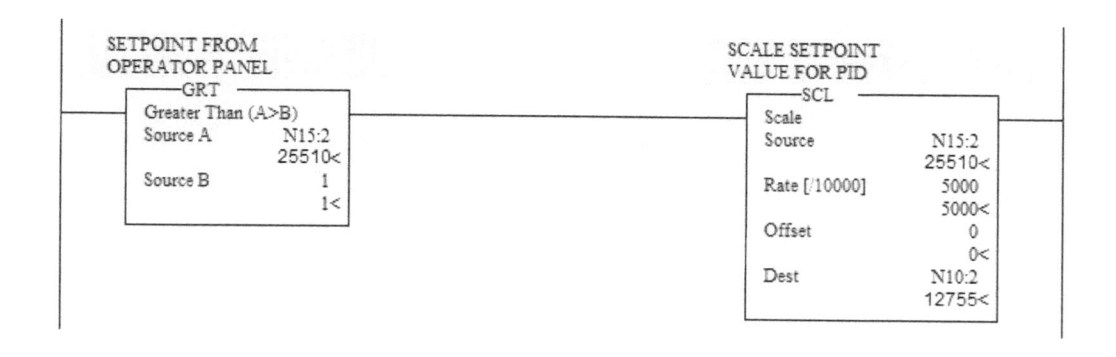

```
    SETPOINT FROM                                SCALE SETPOINT
    OPERATOR PANEL                               VALUE FOR PID
    ─── GRT ───                                  ─── SCL ───
    Greater Than (A>B)                           Scale
    Source A      N15:2                          Source       N15:2
                  25510<                                       25510<
    Source B      1                              Rate [/10000] 5000
                  1<                                           5000<
                                                 Offset        0
                                                               0<
                                                 Dest          N10:2
                                                               12755<
```

In this application, if the setpoint is greater than 1 then the SCL instruction is executed and the input value residing in N15:2 is scaled to a value that corresponds with the range we've set in our setup menu – the **Smax** and **Smin** values contained in words 7 and 8 of the control block. In this case, the setpoint is coming from the HMI or operator interface directly into the N15 file rather than going through any A/D or D/A conversion, such as an input or output module provides. Even though the conversion is digital-to-digital at this point, N15 still requires scaling to fit with the desired range of 0 to 16000.

Remember that when using the SCL instruction, the values that must be entered are the *source* address, the *destination* address (control block word 2), and *the rate* – which is the slope multiplied by 10000, and any *offset* value. This offset value - in our case zero, is also referred to as the y-intercept in Cartesian coordinate terminology. Here is a simple graph showing the scaling conversion.

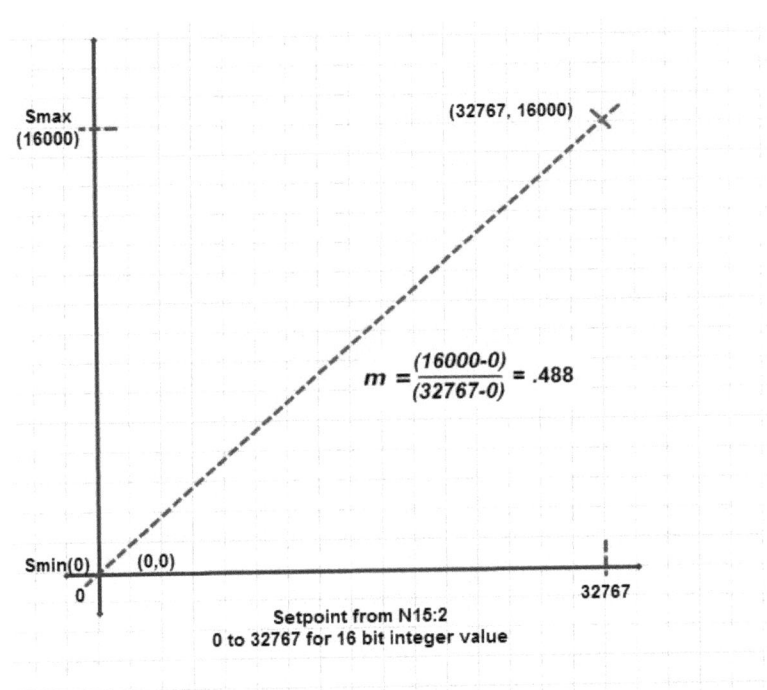

Smax (16000)

(32767, 16000)

$$m = \frac{(16000-0)}{(32767-0)} = .488$$

Smin(0) (0,0)

0   32767

Setpoint from N15:2
0 to 32767 for 16 bit integer value

Using $m = \frac{(y_2-y_1)}{(x_2-x_1)}$ our slope or "rate" value is: .488 which I can round up to .5

This value (.5 multiplied by 10000) is the rate used in our SCL instruction!

Process Variable Scaling:

Our PV input is provided by a 1 to 10 vdc signal and comes into a 1746-NI4 analog input module. This module turns this voltage signal into a corresponding digital integer value that ranges from 0 to 32767. Note that the destination address in the SCL instruction is the address designated in our PID instruction, N11:0.

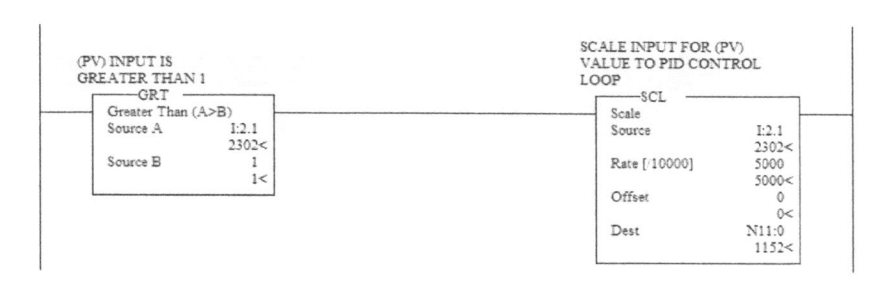

```
                                          SCALE INPUT FOR (PV)
    (PV) INPUT IS                         VALUE TO PID CONTROL
    GREATER THAN 1                        LOOP
    ─── GRT ───                           ─── SCL ───
    Greater Than (A>B)                    Scale
    Source A         I:2.1                Source          I:2.1
                     2302<                                2302<
    Source B         1                    Rate [/10000]   5000
                     1<                                   5000<
                                          Offset          0
                                                          0<
                                          Dest            N11:0
                                                          1152<
```

The range for the process variable and also the control variable is 0 – 16383, so the SCL instruction is set to produce those values.

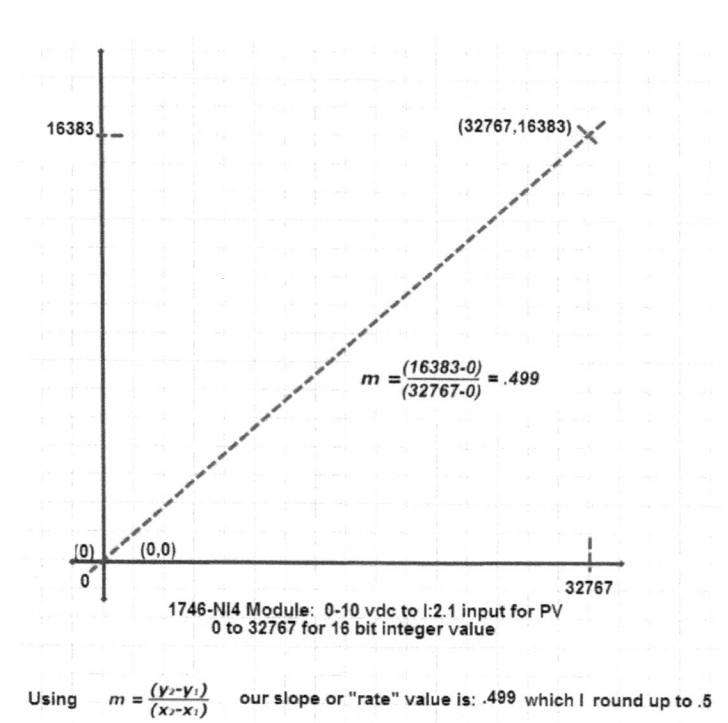

[Graph showing linear relationship from (0,0) to (32767, 16383)]

$$m = \frac{(16383-0)}{(32767-0)} = .499$$

1746-NI4 Module: 0-10 vdc to I:2.1 input for PV
0 to 32767 for 16 bit integer value

Using $m = \frac{(y_2-y_1)}{(x_2-x_1)}$ our slope or "rate" value is: .499 which I round up to .5

This value (.5 multiplied by 10000) is the rate used in our SCL instruction!

Control Variable Scaling:

The scaling for the CV can be a bit tricky, but graphing-it-out is usually helpful in coming up with accurate numbers. First of all, we've selected "yes" on the "Limit Output" option on the setup screen. When selected, the output minimum and maximum values are active – and in this case are set to 30% to 100%. As mentioned earlier, these values are set in words 11 & 12 of the control block.

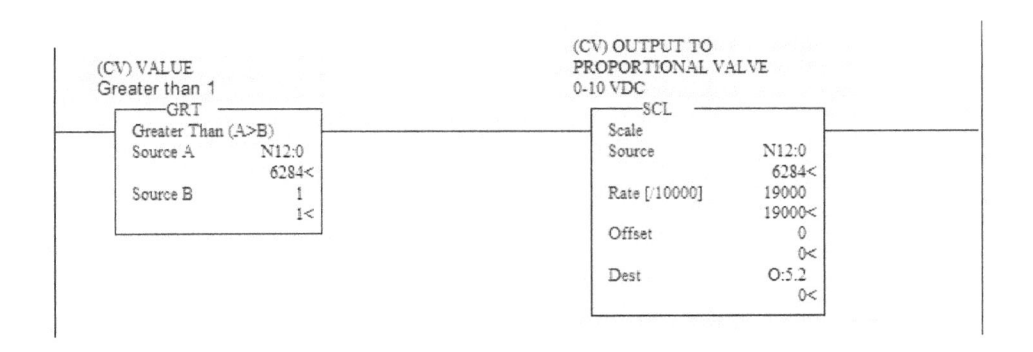

```
                                            (CV) OUTPUT TO
         (CV) VALUE                         PROPORTIONAL VALVE
         Greater than 1                     0-10 VDC
         ─GRT─────────────              ────SCL─────────────
          Greater Than (A>B)               Scale
          Source A      N12:0              Source        N12:0
                        6284<                             6284<
          Source B          1              Rate [/10000] 19000
                           1<                             19000<
                                           Offset            0
                                                             0<
                                           Dest          O:5.2
                                                             0<
```

As you can see on the following graph, the PID produces an output value that will be a maximum numeric value of 16383 – this equates to the 100% value (10 vdc) we want produced by the output module. Therefore, 4915 & 16383 is scaled to 9830 & 32764 respectively – the output module producing 3 vdc at 9830 and 10 vdc at 32764.

The slope (m) or rate for the SCL instruction is calculated using the values shown in the graph and calculates at 1.99 – I used 19000 for the rate setting in this program.

You may also notice that the current output in N12:0 is at 6284, well below our 30% lower limit of 9830. This is the reason that the LL flag bit in word 0 of the control block will set to 1 and will clear during PID execution as the CV rises above 30%.

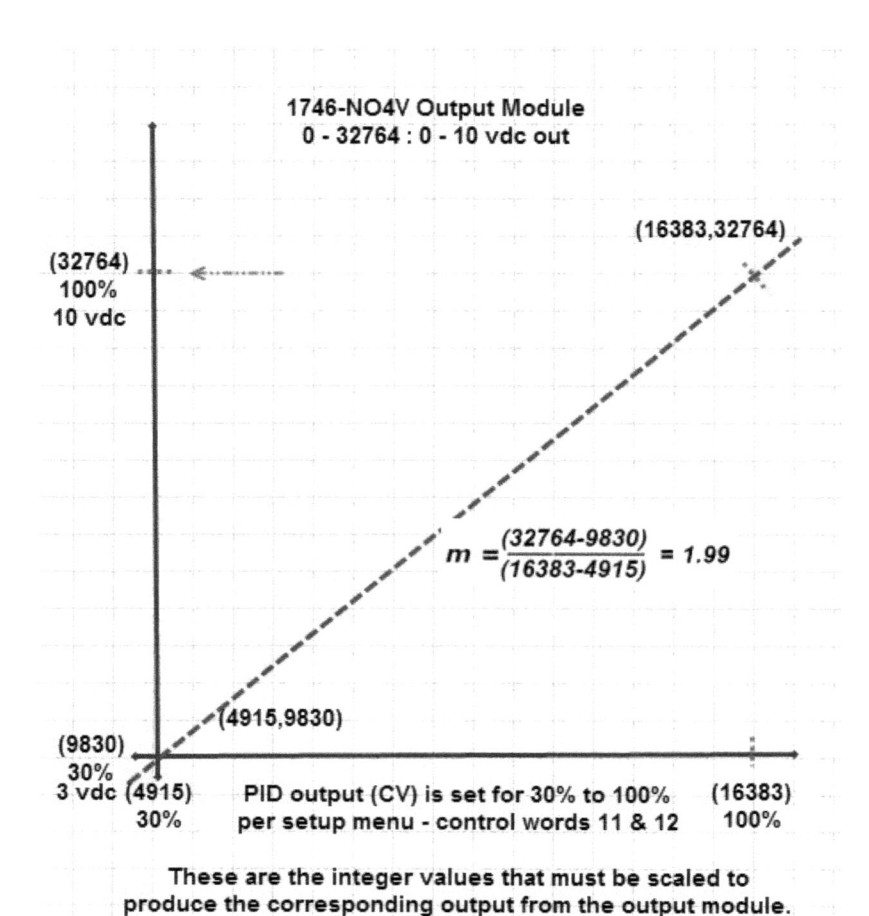

1746-NO4V Output Module
0 - 32764 : 0 - 10 vdc out

(32764)
100%
10 vdc

(16383, 32764)

$$m = \frac{(32764-9830)}{(16383-4915)} = 1.99$$

(4915, 9830)

(9830)
30%
3 vdc (4915)
     30%

PID output (CV) is set for 30% to 100% per setup menu - control words 11 & 12

(16383)
100%

These are the integer values that must be scaled to produce the corresponding output from the output module.

This particular program, as I mentioned earlier, is only using the "P" component of the PID function. This is fine for showing scaling examples and establishing minimum and maximum output values but remember that, in reality, the CV will not look like the linear coordinates shown above. Between limits the rate-of-change of the CV will constantly change in response to the calculated error and the values that integral and derivative functions add the output.

Next, we'll look into the process of "tuning" a PID loop where all three elements; proportional, integral and derivative combine to produce usable output.

Chapter 5: Basic PID Tuning

In this chapter we'll look into the basic method of "tuning" a control loop – as mentioned earlier, often referred to as more "art than science". Of course the object of tuning is to get a control-loop, in this case our PID instruction, to produce an output that will alter our process in the way we desire. For example, we may need production parts to reach a very specific curing temperature within a certain allowed timeframe and then be maintained at that temperature for several hours. The temperature reached and maintained is a critical factor, but so too is the amount of time the process takes to reach the required temperature. Tuning is all about dealing with important factors such as these, and so cannot be ignored even though some process applications can allow for more latitude or be more forgiving than others. Whatever the application - tuning is obviously an important concept to understand.

Here we'll apply and explore some of the general concepts discussed, and get a visual of how different parameter settings perform under the same circumstances. In these examples, and for illustration purposes, I am using a simulator program which provides a good method of keeping the "disturbance" factor of the system at a constant. Remember that due to the dynamic nature of environmental factors that have an effect on any application using PID control, such as heat loss or mechanical restrictions, the settings of a "tuned" control loop will usually vary from recommended starting values. For this reason I consider the recommended starting points for gain, reset and rate values to be just that – a good starting point.

In the following examples, the basic Ziegler-Nichols set-points are used and varied to achieve what I consider optimal performance in controlling a simple heating application. Even though the Ziegler-Nichols settings are different from the RSLogix 500 recommendations, they provide a good example for showing effects on our control and process variables. Here are some definitions for ZN tuning terminology and set-points.

K_C is the "proportional gain" factor already discussed as part of the dependent-gains equation and the gain setting on the PID setup menu.

K_U is the "ultimate gain". This is the K_C setting where consistent oscillation occurs with the process variable (PV) - and for that matter, usually the CV as it attempts to correct the unstable PV. Note that during the step raises of K_C to this point, the T_I and T_D setting are set to zero and the process is subject to proportional control only.

T_U is the oscillation period obtained, also referred to as the **"natural period"**, at the K_U ultimate gain setting. T_U is the time period of one complete cycle – such as a peak-to-peak time approximation.

The suggested initial Ziegler-Nichols settings, when little or no overshoot is desired, are as follows:

- K_C equal to 20% of K_U, (RSLogix 500 manual suggests 50%)
- T_I equal to T_U / 2, (RSLogix 500 manual suggests T_I equal to T_U)
- T_D equal to T_U / 3, (RSLogix 500 manual suggests T_D approx. 1/8 of T_U)

In the following illustration I'm showing a temperature scale with our SP increased to 50 degrees. Note that even though I'm using temperature for an example, this scale could be anything such as 0 to 100% for flow rate, motor speed, etc.

The following illustration shows K_C increased to the point where oscillation, the instability of the PV and CV, is fairly consistent in amplitude and time. This gain setting of approximately 6 is the K_U or *ultimate gain* setting. From this chart we also measure the T_U oscillation period of approximately 80 seconds.

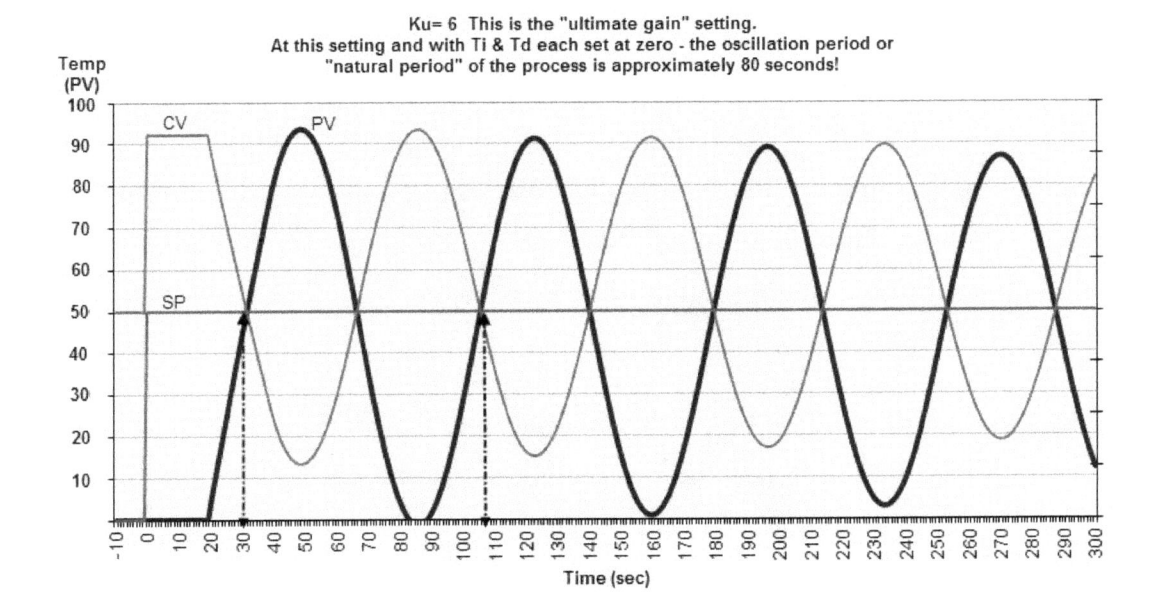

From these approximations we can calculate values to try, at least initially, for K_C, T_I and T_D. The K_C value, set at 20% of K_U, would be approximately 1.35 and is shown in the following illustration. As you can see, we are left with a substantial steady-state error that proportional alone cannot correct.

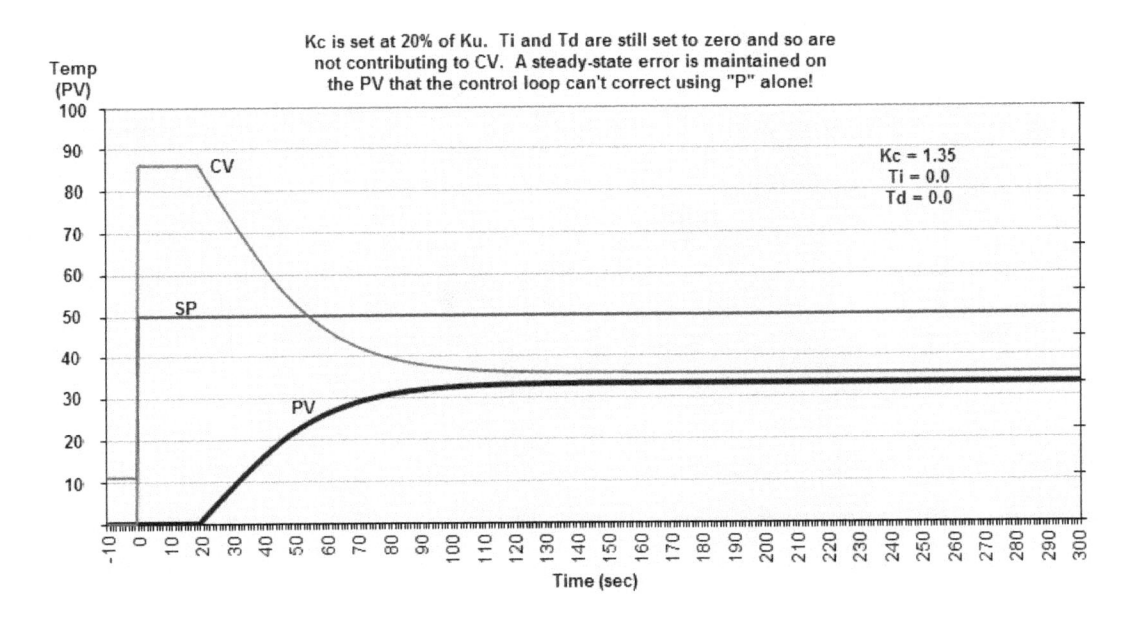

In the next example, the approximated T_I setting of 40 seconds is applied. Recall that with the minutes per repeat algorithm used by the PID instruction, the T_I setting causes the proportional action to be repeated over this 40 second period. As you can see in this illustration, the CV reaches the level of 95% within approximately 12 seconds then begins to slope downward and crossed the SP line at about the 120 second mark. By comparison, in the previous illustration using proportional control alone, the CV didn't reach as high a level and fell off more quickly.

Using P & I alone, with the basic ZN initial settings, allowed the process to reach its SP of 50 degrees but also created an overshoot in the PV of slightly more than 10 degrees. This may or may not be acceptable.

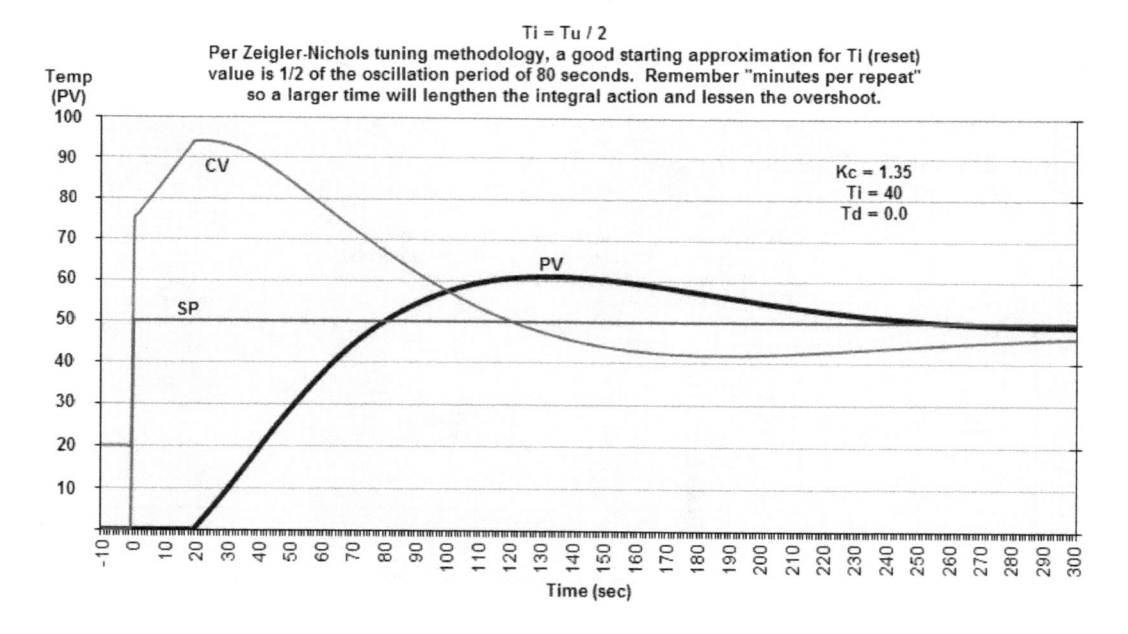

In the next example I've added the derivative component of the Zeigler-Nichols method, the T_D value of approximately 26.6, or (T_U / 3). As you can see, the derivative value, which is our gain value K_C multiplied by rate-of-change of our error rather than error directly, dampens but doesn't entirely eliminate the overshoot.

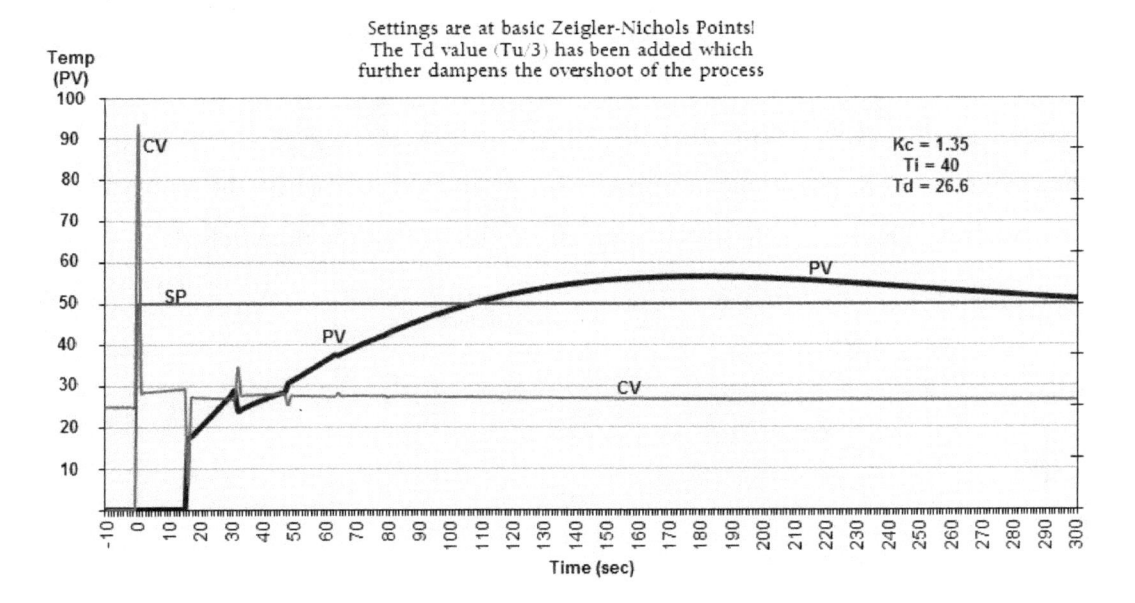

[Graph: Settings are at basic Zeigler-Nichols Points! The Td value (Tu/3) has been added which further dampens the overshoot of the process. Kc = 1.35, Ti = 40, Td = 26.6]

Not desiring the overshoot or the length of time it has taken the PV to settle back to the SP, I decided to try something different. Here I've set T_I and T_D back to zero, and increased the K_C value to 3.0, or 50% of K_U. This still left some of the steady-state error, although not as much.

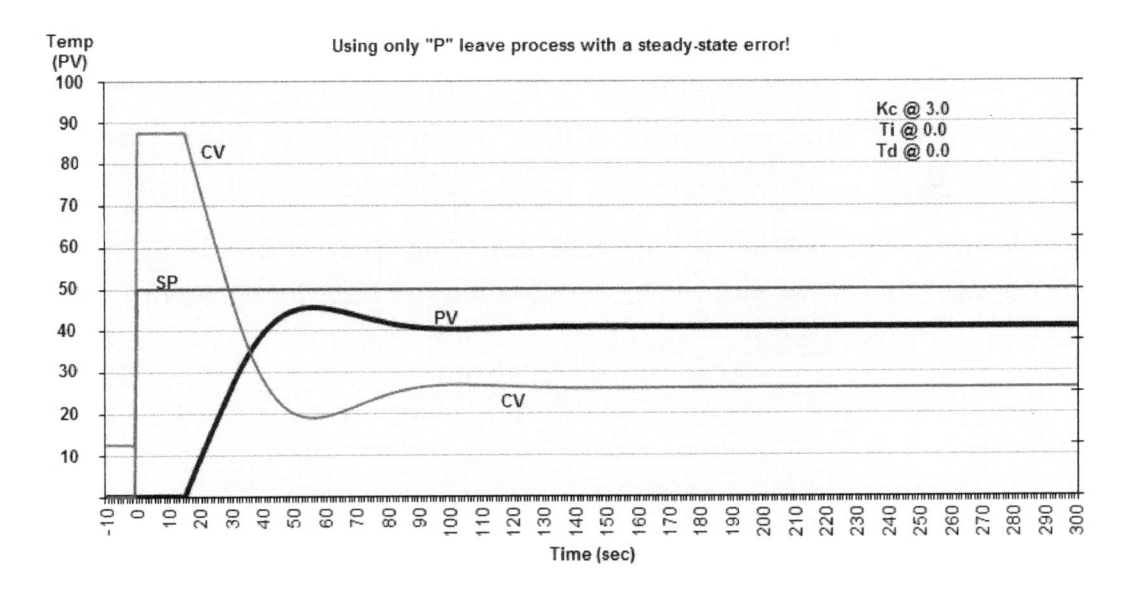

[Graph: Using only "P" leave process with a steady-state error! Kc @ 3.0, Ti @ 0.0, Td @ 0.0]

What is more interesting is what happens when the same calculated T_I and T_D values are put back into the algorithm. The next illustration now shows only slight overshoot using the same T_I value of 40. Note that the integral action, the repeat of the proportional action, makes a more pronounced impact on CV and the PV when the gain value has been increased. The PV reaches SP in a much shorter timeframe than before.

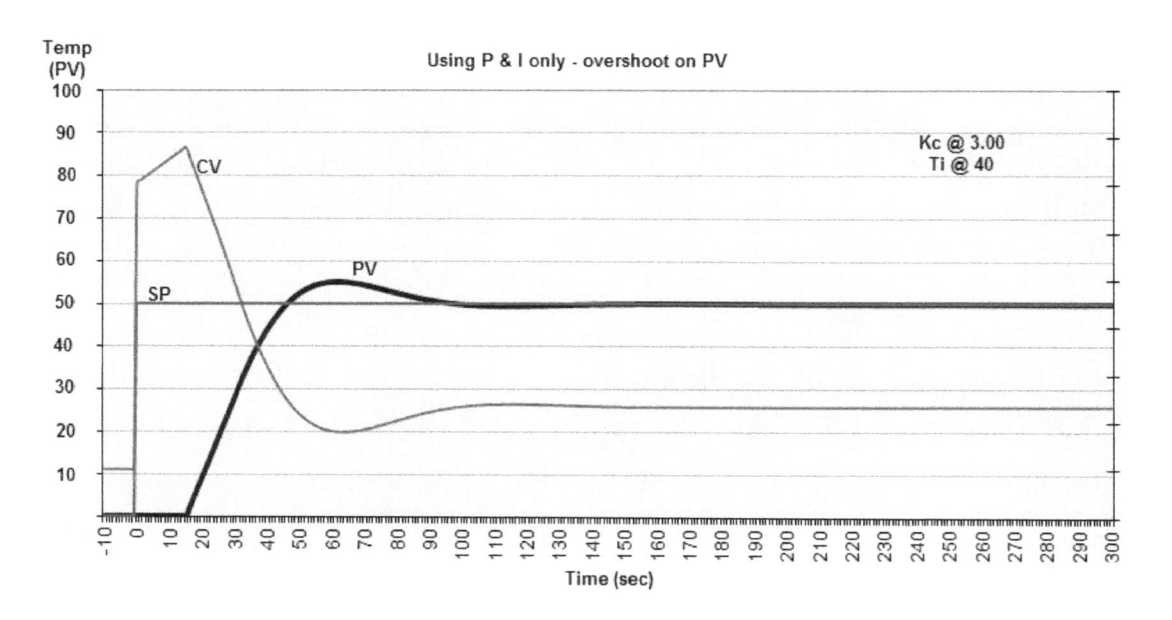

When the derivative value of 26.6 is put back into the instruction algorithm some dampening effect is introduced which yields the following process curve. This curve shows no overshoot while bringing the PV to its setpoint in an appropriate time interval and the integral action keeps the PV tracking right along the SP as time progresses. All things considered – a better tuned control process than we had with the K_C of 1.35 used at the beginning of the example.

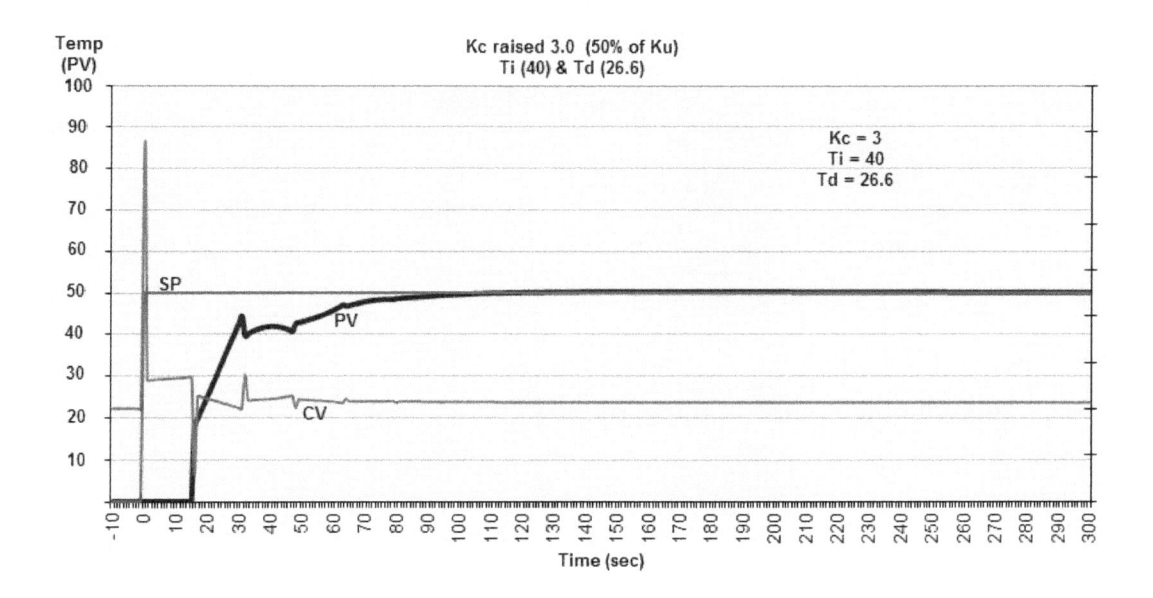

Other important details to remember when setting up PID instructions in a ladder program, is the *Loop Update Time* and the *Time Mode* settings. To review, the Loop Update Time determines the time interval when the program will execute the PID instruction. The time interval range is from 1 to 1024 in 0.01 second intervals. This time should be set at a value five to ten times <u>faster</u> than the natural period of the output – what the ZN method refers to as the T_U value. So in this example the Loop Update Time could, hypothetically, be set to 800 which would set the time interval to 8 seconds. The application would also be set for "timed" mode rather than STI because the natural period of this simulated process is greater than 10 times the scan time of the processor.

What I've shown in this example is referred to as the Ziegler-Nichols "Continuous Tuning" method using the calculation values that produce no overshoot. Some of the calculation values listed with variants of this method allows overshoot with decaying oscillations. Once again, with some applications this is not an important issue. My work experience has mostly been with robotics and CNC machines - so any overshoot seems like a critical issue to me and something to be avoided if possible.

Here is a recap of the steps involved with this basic tuning method along with the chart showing the calculation values for the Ziegler-Nichols method.

- Step 1: Place control in "manual mode".
- Step 2: Adjust SP to a desired "safe" level. (overshoot will not cause damage)
- Step 3: Set T_I and T_D to zero values.
- Step 4: Set K_C to zero.
- Step 5: Switch to "Auto" mode
- Step 6: Increase SP by 10% step increases until a sustained and continuous oscillation occurs and is maintained.
- Step 7: Record T_U "ultimate period" and K_U "ultimate gain" values.
- Step 8: Use chart values to set K_C, T_I and T_D values if needed.
- Step 9: Test settings in automatic mode.
- Step 10: Re-adjust as necessary.

Ziegler-Nichols
Continuous Tuning Method Setpoints

Control Options	(Kc) gain	(Ti) reset	(Td) rate
P	0.5 Ku	NA	NA
P I	0.45 Ku	Tu /1.2	NA
PD	0.8 Ku	NA	Tu /8
PID	0.60 Ku	Tu /2	Tu /8
Some overshoot	0.33 Ku	Tu /2	Tu /3
No overshoot	0.2 Ku	Tu /2	Tu /3

When using the Ziegler-Nichols method, remember that it is by definition a "heuristic technique", so initial results can be inexact or approximate. This method can however, get a control-loop operational to the point where other changes can then be entered – just as was done in the previous example until a desired process curve is achieved.

Chapter 6: Concluding Comments

Whether you are routinely called upon to program PID controls for new projects, or simply to troubleshoot existing machine and process applications – a good working familiarity of control-loop technology must be a part of a technician's knowledge base. More and more, if programming new projects, you probably find yourself using RSLogix 5000 software and ControlLogix or CompactLogix processors. These applications will of course give you more options for PID control, the type of algorithm used and how tuning can be accomplished. However there are still plenty of SLC and Micrologix processors in use and endless applications where RSLogix 500 software can be effectively utilized. Apart from these facts, I think RSLogix 500 is great development software with which to learn programming techniques – including basic PID concepts. I hope you feel, at this point, somewhat comfortable with the operation and tuning of any controller – whether PLC based or not. If so, then this short book will have accomplished its basic goal.

Some of the topics and objectives covered in this book have been:

- The basic PID control loop algorithm and what it does.
- Programming and set-up of the PID instruction.
- PID control terminology and definitions.
- The PID control block - word elements and discrete bits.
- Selecting a Time Mode: STI or Timed.
- Setting the Loop Update Time interval.
- Establishing CV limits.
- Scaling techniques for PID input, output and setpoint values.
- Ziegler-Nichols terminology and variables.
- Ziegler-Nichols Continuous Tuning Method.

As with the other books I've written that pertain to programming, motor control, industrial networking and VFD drive installation, the goal of this book has been to convey these topics, as well as possible, from the written page - along with some useful illustrations and examples. My experience with electrical and electronics troubleshooting, as well as PLC programming, is that nothing is more effective in helping one acquire new skills than "jumping in with both feet". Find an application which can benefit by implementing PLC control and go to work on the problem. Take notes as you work through different problems that arise. At the end of the project you will usually be surprised at the proficiency and skills you've gained.

To that end I wish you the very best!

***

Other Books by Gary D. Anderson

Practical Guides for the Industrial Technician:

- *Motion Control for CNC & Robotics*
- *Variable Frequency Drives – Installation & Troubleshooting*
- *Industrial Network Basics*

RSLogix 500 Programming Series:

- *Basics Concepts of Ladder Logic Programming*
- *Advanced Programming Concepts*
- *Ladder Logic Diagnostics & Troubleshooting*
- *PID Programming Using RS Logix 500*
- *Program Flow Instructions Using RS Logix 500*

RSLogix 5000 & ControlLogix Controllers:

- *RSLogix 5000 - Understanding ControlLogix Basics*: this new book covers the many essential details and concepts that serve as building blocks for sucessful programming and troubleshooting within the ControlLogix platform.

As I've said before, I know you have many options when choosing from books and online resources that discuss these topics; so I thank you for selecting my book - I hope you feel you've benefited by doing so. As always, I welcome your comments and feedback. My goal is to present, in clear and concise language, relevant technical topics, and I have found feedback from my readers to be an invaluable resource. With that said, if you would like to contact me with questions or comments you can do so at the following email address:

Email: ganderson61@cox.net

Your reviews on Amazon are helpful and appreciated. If you've enjoyed what you've read and feel this book has provided a positive benefit to you, then please take a moment and write a short review.